Preface

Welcome to the "Science Olympiad All Chapters Combined MCQ Class 1" – a dynamic resource designed to spark curiosity and foster a deep understanding of science in young minds. This book serves as a comprehensive guide for students preparing for Science Olympiad examinations, covering every chapter of the Class 1 science syllabus. It invites learners to explore the wonders of the world through engaging multiple-choice questions that are both thought-provoking and aligned with the exam format, setting a solid foundation for a lifelong journey of scientific inquiry.

Crafted with care, this book goes beyond mere test preparation by nurturing critical thinking, logical reasoning, and problem-solving skills. Each question is carefully selected to be age-appropriate and challenging, while detailed answers and explanations support independent learning and reinforce key concepts. With its holistic approach to skill development and exam readiness, this resource is poised to become an invaluable companion for every Class 1 student, inspiring a genuine love for science and the thrill of discovery.

Table of Contents

Preface 1

Chapter 1: MCQs Part A 3

Chapter 2: MCQs Part B 13

Chapter 3: MCQs Part C 23

Chapter 4: MCQs Part D 32

Chapter 1: MCQs Part A

1. What do plants need to grow?

a) Candy

b) Water

c) Soda

d) Ice cream

Ans:- b) Water

2. Which part of the plant grows underground?

a) Leaf

b) Flower

c) Root

d) Stem

Ans:- c) Root

3. What is the color of the sky on a clear day?

a) Red

b) Blue

c) Green

d) Yellow

Ans:- b) Blue

4. Which one is a source of light?

a) Moon

b) Sun

c) Star

d) All of the above

Ans:- d) All of the above

5. Which animal is known as the king of the jungle?

a) Elephant

b) Tiger

c) Lion

d) Bear

Ans:- c) Lion

6. How many legs does a spider have?

a) Six

b) Eight

c) Ten

d) Twelve

Ans:- b) Eight

7. What do bees make?

a) Milk

b) Honey

c) Cheese

d) Butter

Ans:- b) Honey

8. What do we use to see tiny objects?

a) Microscope

b) Telescope

c) Binoculars

d) Magnifying glass

Ans:- a) Microscope

9. Which planet is known as the Red Planet?

a) Earth

b) Mars

c) Venus

d) Jupiter

Ans:- b) Mars

10. Which sense organ helps us to hear?

a) Eyes

b) Nose

c) Ears

d) Tongue

Ans:- c) Ears

11. What is the baby of a dog called?

a) Kitten

b) Puppy

c) Calf

d) Chick

Ans:- b) Puppy

12. Which of the following can fly?

a) Fish

b) Dog

c) Bird

d) Rabbit

Ans:- c) Bird

13. How many colors are there in a rainbow?

a) Five

b) Six

c) Seven

d) Eight

Ans:- c) Seven

14. What do we use to brush our teeth?

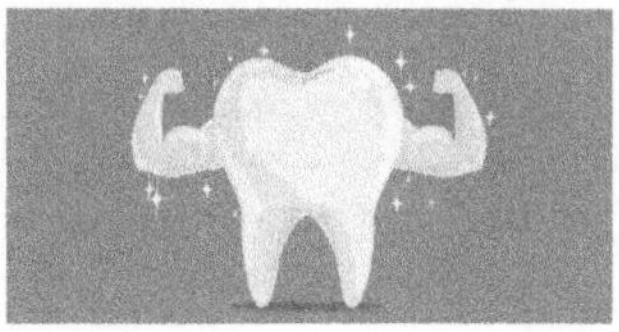

a) Comb

b) Brush

c) Toothbrush

d) Spoon

Ans:- c) Toothbrush

15. Which fruit is yellow in color and monkeys love to eat?

a) Apple

b) Banana

c) Grape

d) Orange

Ans:- b) Banana

16. What covers our body and keeps us warm?

a) Skin

b) Clothes

c) Blanket

d) Hair

Ans:- b) Clothes

17. Which part of the plant is green and makes food for the plant?

a) Root

b) Stem

c) Leaf

d) Flower

Ans:- c) Leaf

18. What is the largest mammal?

a) Elephant

b) Blue Whale

c) Giraffe

d) Hippopotamus

Ans:- b) Blue Whale

19. What do birds use to build their nests?

a) Stones

b) Mud

c) Twigs and leaves

d) Sand

Ans:- c) Twigs and leaves

20. What is the shape of the Earth?

a) Flat

b) Square

c) Round

d) Triangular

Ans:- c) Round

21. What do cows give us?

a) Eggs

b) Wool

c) Milk

d) Bread

Ans:- c) Milk

22. Which one of these is a fruit?

a) Carrot

b) Potato

c) Tomato

d) Cabbage

Ans:- c) Tomato

23. What do we breathe in to live?

a) Smoke

b) Dust

c) Air

d) Water

Ans:- c) Air

24. How many seasons are there in a year?

a) Two

b) Three

c) Four

d) Five

Ans:- c) Four

25. What do you call the process of plants making food using sunlight?

a) Digestion

b) Photosynthesis

c) Evaporation

d) Condensation

Ans:- b) Photosynthesis

Chapter 2: MCQs Part B

1. What is the sound made by a cat called?

a) Barking

b) Meowing

c) Roaring

d) Chirping

Ans:- b) Meowing

2. Which part of our body helps us to smell?

a) Eyes

b) Nose

c) Ears

d) Hands

Ans:- b) Nose

3. Which of these is not an insect?

a) Butterfly

b) Ant

c) Spider

d) Snake

Ans:- d) Snake

4. What do you call a young horse?

a) Puppy

b) Kitten

c) Foal

d) Cub

Ans:- c) Foal

5. What do humans need to drink every day to stay healthy?

a) Juice

b) Milk

c) Water

d) Soda

Ans:- c) Water

6. Which of these animals live in water?

a) Dog

b) Fish

c) Cat

d) Rabbit

Ans:- b) Fish

7. What color are leaves usually?

a) Red

b) Blue

c) Green

d) Yellow

Ans:- c) Green

8. What do we use to cut paper?

a) Pencil

b) Eraser

c) Scissors

d) Ruler

Ans:- c) Scissors

9. What is the main gas in the air we breathe?

a) Oxygen

b) Nitrogen

c) Carbon dioxide

d) Hydrogen

Ans:- b) Nitrogen

10. What do frogs do to catch insects?

a) Bark

b) Chirp

c) Jump

d) Extend their tongue

Ans:- d) Extend their tongue

11. Which of these is a domestic animal?

a) Lion

b) Cow

c) Wolf

d) Tiger

Ans:- b) Cow

12. Which sense organ helps us to taste?

a) Eyes

b) Nose

c) Tongue

d) Ears

Ans:- c) Tongue

13. Which is the tallest animal?

a) Elephant

b) Giraffe

c) Zebra

d) Lion

Ans:- b) Giraffe

14. Which of these is not a living thing?

a) Tree

b) Dog

c) Stone

d) Bird

Ans:- c) Stone

15. What is the color of an apple?

a) Blue

b) Green

c) Red

d) Yellow

Ans:- c) Red

16. What helps fish to swim in water?

a) Legs

b) Fins

c) Wings

d) Tail

Ans:- b) Fins

17. What do we use to write on a blackboard?

a) Chalk

b) Pen

c) Pencil

d) Crayon

Ans:- a) Chalk

18. What do you call water in its solid form?

a) Steam

b) Ice

c) Vapor

d) Fog

Ans:- b) Ice

19. Which bird is known for its colorful feathers?

a) Crow

b) Peacock

c) Sparrow

d) Owl

Ans:- b) Peacock

20. What do you call a baby sheep?

a) Calf

b) Lamb

c) Kid

d) Cub

Ans:- b) Lamb

21. Which part of the body helps us to see?

a) Ears

b) Nose

c) Eyes

d) Tongue

Ans:- c) Eyes

22. Which of these can we drink?

a) Water

b) Sand

c) Rock

d) Plastic

Ans:- a) Water

23. Which season is known for snowfall?

a) Summer

b) Winter

c) Autumn

d) Spring

Ans:- b) Winter

24. What do hens lay?

a) Eggs

b) Milk

c) Wool

d) Cheese

Ans:- a) Eggs

25. Which of these is a reptile?

a) Dog

b) Snake

c) Cat

d) Elephant

Ans:- b) Snake

Chapter 3: MCQs Part C

1. What is the color of grass?

    a) Red

    b) Green

    c) Yellow

    d) Blue

Ans:- b) Green

2. What do we use to write on paper?

    a) Spoon

    b) Chalk

    c) Pen

    d) Plate

Ans:- c) Pen

3. What helps birds to fly?

    a) Legs

    b) Beaks

    c) Wings

    d) Tails

Ans:- c) Wings

4. Which of these is a natural source of light?

   a) Candle

   b) Bulb

   c) Sun

   d) Torch

Ans:- c) Sun

5. Which part of our body helps us to hold things?

   a) Feet

   b) Hands

   c) Ears

   d) Nose

Ans:- b) Hands

6. Which of these animals can hop?

   a) Elephant

   b) Frog

   c) Dog

   d) Cat

Ans:- b) Frog

7. What is the color of the sun?

a) Blue

b) Green

c) Yellow

d) Red

Ans:- c) Yellow

8. What is the outer covering of a tree called?

a) Bark

b) Leaf

c) Flower

d) Root

Ans:- a) Bark

9. Which of these can we eat?

a) Stone

b) Apple

c) Paper

d) Plastic

Ans:- b) Apple

10. Which animal has a trunk?

a) Lion

b) Elephant

c) Tiger

d) Monkey

Ans:- b) Elephant

11. What do you call the sound made by a cow?

a) Baa

b) Moo

c) Meow

d) Quack

Ans:- b) Moo

12. Which part of the body helps us to walk?

a) Hands

b) Feet

c) Nose

d) Ears

Ans:- b) Feet

13. Which of these is a carnivore?

a) Rabbit

b) Deer

c) Lion

d) Cow

Ans:- c) Lion

14. What do you call a baby cat?

a) Puppy

b) Cub

c) Kitten

d) Chick

Ans:- c) Kitten

15. Which of these is a source of water?

a) River

b) Car

c) House

d) Tree

Ans:- a) River

16. Which part of the plant is colorful and attracts insects?

a) Root

b) Stem

c) Leaf

d) Flower

Ans:- d) Flower

17. What do you call the process of water turning into vapor?

a) Condensation

b) Evaporation

c) Precipitation

d) Freezing

Ans:- b) Evaporation

18. Which of these is a herbivore?

a) Lion

b) Cow

c) Wolf

d) Snake

Ans:- b) Cow

19. What do we use to measure time?

a) Scale

b) Clock

c) Compass

d) Thermometer

Ans:- b) Clock

20. What do you call a group of stars that form a pattern?

a) Planet

b) Galaxy

c) Constellation

d) Meteor

Ans:- c) Constellation

21. Which of these animals has feathers?

a) Fish

b) Bird

c) Dog

d) Elephant

Ans:- b) Bird

22. What do we use to keep our body clean?

a) Shampoo

b) Soap

c) Detergent

d) Toothpaste

Ans:- b) Soap

23. Which of these is a source of energy?

a) Sun

b) Paper

c) Rock

d) Plastic

Ans:- a) Sun

24. Which part of the plant makes seeds?

a) Root

b) Stem

c) Leaf

d) Flower

Ans:- d) Flower

25. Which animal is known for its black and white stripes?

a) Lion

b) Tiger

c) Zebra

d) Leopard

Ans:- c) Zebra

Chapter 4: MCQs Part D

1. Which of these is not a planet?

   a) Mars

   b) Earth

   c) Sun

   d) Venus

Ans:- c) Sun

2. What do you call the change of water from liquid to solid?

   a) Melting

   b) Freezing

   c) Boiling

   d) Evaporation

Ans:- b) Freezing

3. Which of these is a vegetable?

   a) Apple

   b) Carrot

   c) Grape

   d) Banana

Ans:- b) Carrot

4. What is the color of the sun during sunset?

a) Green

b) Blue

c) Orange

d) Purple

Ans:- c) Orange

5. Which animal lives in a den?

a) Fish

b) Bird

c) Rabbit

d) Lion

Ans:- d) Lion

6. What is the main source of energy for plants?

a) Water

b) Soil

c) Sunlight

d) Wind

Ans:- c) Sunlight

7. Which of these animals is known for its long neck?

a) Elephant

b) Giraffe

c) Zebra

d) Lion

Ans:- b) Giraffe

8. What do we use to keep our hair tidy?

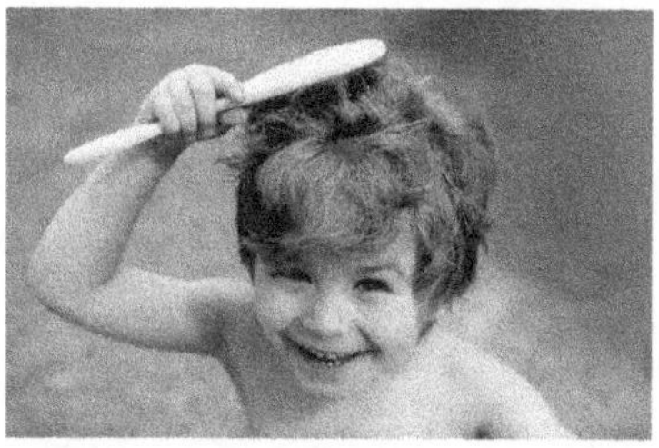

a) Brush

b) Comb

c) Both a and b

d) Spoon

Ans:- c) Both a and b

9. Which of these is a farm animal?

a) Shark

b) Cow

c) Elephant

d) Tiger

Ans:- b) Cow

10. What do you call a young chicken?

a) Kitten

b) Puppy

c) Chick

d) Calf

Ans:- c) Chick

11. Which of these is a liquid?

a) Ice

b) Water

c) Steam

d) Rock

Ans:- b) Water

12. Which of these animals can swim?

a) Cat

b) Dog

c) Fish

d) Rabbit

Ans:- c) Fish

13. What is the color of the moon?

a) Red

b) Blue

c) White

d) Green

Ans:- c) White

14. Which of these is a tool used by a doctor?

a) Stethoscope

b) Hammer

c) Saw

d) Wrench

Ans:- a) Stethoscope

15. What is the color of coal?

a) Red

b) Yellow

c) Black

d) Green

Ans:- c) Black

16. Which of these is a part of the human body?

a) Engine

b) Leaf

c) Arm

d) Wing

Ans:- c) Arm

17. Which of these animals is known for spinning webs?

a) Bee

b) Butterfly

c) Spider

d) Ant

Ans:- c) Spider

18. What do you call the movement of the Earth around the Sun?

a) Rotation

b) Revolution

c) Spinning

d) Circulation

Ans:- b) Revolution

19. Which of these is not a mammal?

a) Dog

b) Fish

c) Cat

d) Cow

Ans:- b) Fish

20. What is the main source of light on Earth?

a) Moon

b) Sun

c) Stars

d) Lamps

Ans:- b) Sun

21. Which of these is a mode of transport?

a) Tree

b) Car

c) Rock

d) River

Ans:- b) Car

22. Which of these is a type of weather?

a) Sunny

b) Loud

c) Heavy

d) Bright

Ans:- a) Sunny

23. What do we call the smallest unit of life?

a) Atom

b) Cell

c) Molecule

d) Particle

Ans:- b) Cell

24. Which of these is a common pet?

a) Elephant

b) Tiger

c) Cat

d) Whale

Ans:- c) Cat

25. What do we wear to protect our eyes from the sun?

a) Hat

b) Sunglasses

c) Gloves

d) Scarf

Ans:- b) Sunglasses

9 789369 728855

Printed at Repro India Ltd.